CW00497581

The Tiger Tank Story

The Tiger Tank Story

Mark Healy

The
History
Press

Published in the United Kingdom in 2010 by
The History Press
The Mill · Brimscombe Port · Stroud · Gloucestershire · GL5 2QG

Copyright © The History Press 2010

British Library Cataloguing in Publication Data
A catalogue record for this book is available from the British
Library.

Hardback ISBN 978-0-7524-5629-4

Half title page: *Tiger I of s.Pz.Abt 503, Eastern Front, January 1943.*

Title page: *Tiger II of s.Pz.Abt 506, the Ardennes campaign, December 1944.*

Typesetting and origination by The History Press

CONTENTS

Tiger II of Panzerabteilung 316 (Fkl), France, June 1944.

INTRODUCTION

This is a small book about a very big subject. The Tiger tank – the most famous tank ever built – has been the focus of some detailed and very comprehensive studies by a number of authors. Foremost among them is the three-volume set on the Tiger I and II by Jentz and Doyle and the two volumes on the Tiger in Action by Wolfgang Schneider. This book is in no way attempting to emulate their exhaustive treatment of the subject or others like it. It contains text, pictures, drawing and colour profiles to allow those new to this subject to get a sense of the fascination these German beasts still exert on historians, enthusiasts and modellers.

As in many other subjects to do with the Second World War, there is currently more information available than ever. It is also now possible to see a Tiger I and Tiger II moving in the 'metal' so to speak. After many years of time consuming restoration an early production Tiger I that was captured by the British Army in Tunisia in 1943 has been restored to working order at the Tank Museum at Bovington Camp in Dorset, England. Similarly there is a Tiger II in running order at the French Army Tank Museum at Saumur. To witness an outing of these machines on one of the public days held in the summer is well worth the effort. While they are now nothing more than museum pieces, their guns forever silent, it does not take much imagination to appreciate the fear that these machines generated on the many battlefields on which they served between 1943 and 1945.

I would like to thank the following people for their very kind help in providing me with pictures, profiles and colour plates. Two big thank yous to Alan Hamby in the States and George Bradford in Canada. They have made my life so much easier with their help. Alan Hamby runs an excellent website called the 'Tiger I Information Centre' on www.alanhamby.com. Alan very kindly allowed me to use images and information from his site. George Bradford also kindly permitted to use his profile drawings of the different variants of the Tigers I and II. They allowed me to save many words in a short text. George is the founder of the excellent long running 'AFV News'. This a publication well worth subscribing to. Finally, the gentlemen at the Polish company of *Wydawnictwo Militaria*, for the use of their excellent colour profiles.

The combat debut of what was to emerge as the most famous tank of the Second World War could hardly have been more inauspicious. On the orders of Adolf Hitler, four Tiger I tanks of heavy tank battalion 502 were committed to action on 22 September 1942 in support of German 170th Infantry Division operating to the south of Leningrad (now St.Petersburg) in Russia. Ever anxious to see how new equipment performed and believing the heavy tank to be a battle winner for the *Panzerwaffe,* the German leader had chosen that it be used in forested and marshy terrain totally unsuited to its strengths. In consequence, Soviet anti-tank guns, well camouflaged amidst the dense vegetation, knocked out three Tigers while the fourth became stuck in marshy soil. While the former were recovered, the latter was so mired in the swamp that it had to be abandoned. There was much consternation that it would be captured by the Russians and although it was ordered to be recovered, it was only towards the end of November that the hulk of the machine, stripped of all recoverable equipment, was blown up. Despite this blip, the appearance of the Tiger I elsewhere on the Eastern Front and in Tunisia in the months to come would lead to the fostering of an almost mythic reputation for the effectiveness of this heavy panzer and its stable mate and successor, the Tiger II.

The genesis of a new heavy tank that would eventually emerge as the Tiger I stemmed from a meeting held by Hitler with members of the army and German industry on 26 May 1941, although its heritage can be traced back to 1937. In light of the encounter of the panzers with heavily armoured allied tank designs such as the French Char 1B and the British Matilda II in the French Campaign of 1940, and assumptions being made about future enemy tank developments, the German leader stressed the need for a machine with a gun of superior penetrative power, carrying much heavier armour and a maximum speed of at least 40kmph. The first of these requirements he saw as being satisfied by the adaptation of the 88mm anti-aircraft gun. This would permit the new tank to defeat one 100mm of armour at 1400–1500m. Protection would be provided by frontal armour of 100mm and 60mm on the flanks.

Just a month before, Dr Ferdinand Porsche – who had been working independently on his own heavy tank project in the 45-ton class – had come to an agreement with the gun manufacturer Krupp to design a turret equipped with a variant of the 88mm Flak 18 anti-aircraft gun adapted for use on his design. However, at his meeting on 26 May Hitler let it be known that he wished to see a variant of the new 88mm Flak 41 fitted to the turret being built by Krupp for the Porsche heavy tank. The Flak 41 was a bigger and more powerful weapon than the Flak 18, and it was quickly ascertained that it was too large to fit inside the Krupp turret. Nonetheless, the desire to employ this weapon in a tank would lead to the development of an even heavier heavy panzer – the Tiger II.

Hitler ordered that the Porsche design, as well as that of Henschel of Kassel, be built alongside one another with six examples of both machines being available for comparative trials by the summer of 1942. Henschel had not up to this point been involved in producing a heavy tank design. Their own project, proceeding under the designation VK36.01 (VK = *Volket-tenfahrzeuge* or 'fully tracked experimental vehicle', the number 36 and 45 denoting the weight class of the tank with 01 indicating that it was first design to the specification) had been to develop a new tank in the

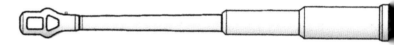

Did you know?

Dr Ferdinand Porsche was also responsible for designing a car known to millions as the Volkswagen Beetle. In 1933, Hitler invited Porsche to submit designs to him for a car that could purchased 'by the people'. The protoypes were running by 1936 and the new car was put into production in 1939. However, almost none were produced for civilian use before the war. The best known variant was the Kubelwagen – the German Army equivalent of the US Jeep.

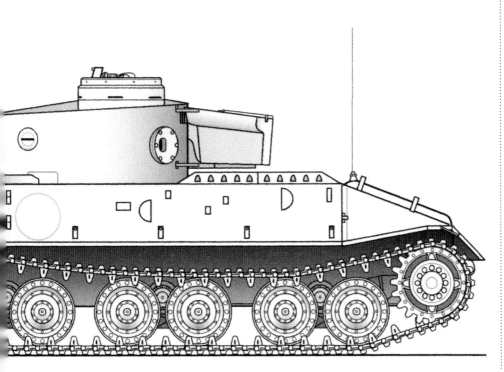

◄ *The prototype of the unsuccessful Porsche Tiger.*

36-ton class. It had been intended that this would be armed with a new 75mm gun. The project was halted just two days after the conference when Henschel received an instruction to modify their medium tank chassis to mount a turret with an 88mm gun. This required Henschel to enlarge the hull of the VK36.01 to accommodate the Krupp turret. This was done by adding sponsons over the tracks and adding a third set of outer road wheels to shoulder the increased weight, which would rise, when eventually employed in combat, to 56 tonnes. The competing machines were both allocated the designation of VK45.01 with the letters (P) and (H) to indicate their company origins under the generic title of the 'Tiger program'.

Although the commitment to produce a heavy tank was taken before the launch of Operation Barbarossa – the German invasion of the Soviet Union on 22 June 1941 – events on the Eastern Front forced the rapid development of such a machine so as to combat the appearance of a new generation of superior Soviet tanks that had rendered the primary German panzers – in the Marks III and IV – technically obsolescent. The shock generated by the Soviet T-34 and KV-1, of whose existence German military intelligence had not been aware, was profound.

The T-34 presented a masterful balance of thick, sloped armour, heavy armament and exceptional all-terrain mobility in all weathers, whereas the KV-I – weighing in at over 50 tons and mounting even heavier armour than the T-34, but the same gun – was impervious to all extant German tank-mounted and anti-tank weapons. Yet

again, recourse to employment of the 88mm Flak guns in the anti-tank role in the field was often the only means available to defeat these enemy tanks. Luckily for the Germans, in the opening months of the Russian campaign their operational technique more than compensated in the face of superior enemy tanks whose poorly trained crews were unable to exploit the potential of their new charges. Furthermore, the KV tank in particular was also suffering from reliability problems. Nonetheless, the prognosis for the Wehrmacht and especially the *Panzerwaffe,* once it became clear that the Red Army would not be defeated quickly as had been presumed by all of their pre-invasion planning, was that these Soviet tanks would be encountered in ever greater numbers. Something clearly had to be done!

A fact-finding visit in November 1941 by representatives from the army weapons department and the armaments industry overseeing tank design and production to examine these Soviet designs led to the decision to develop a new medium tank – the Panther – and to expedite development and production of the new heavy tank and get it into production and service as soon as possible.

By dint of great effort, both companies had prototypes of their respective machines on hand at Rastenburg – the site of Hitler's Eastern Front headquarters in East Prussia – for inspection and demonstration for his birthday on 20 April 1942. They were put through their paces with mixed fortunes. Although sentimentally inclined towards the Porsche tank as its genial designer was one of his personal favourites, it was the nonetheless the Henschel machine that Hitler sanctioned for production. The Achilles heel of the Porsche design was its petrol-electric motors – one powering each track. They proved unreliable and over-complex and the machine was deemed unsuitable for combat employment as a tank. Despite this decision, production of

90 Porsche Tigers had already begun, so certain had the Herr Doktor been of the selection of his machine. Having lost the competition, production halted. The already completed chassis were later converted to heavy assault guns mounting the Pak 43 – the tank version of the Flak 43. Named Ferdinand in honour of its designer, it first saw service during Operation Citadel in July 1943.

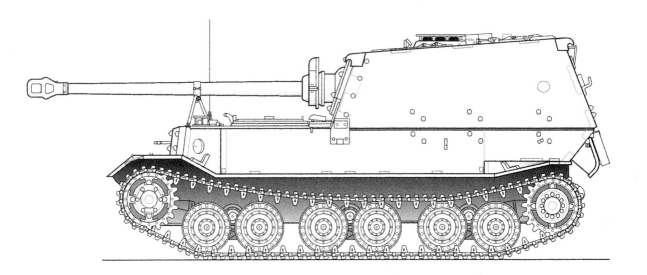

The winning design received its official designation as Tiger I, Model H/E, VK45.01(H) – sd.Kfz.181 and was committed to production at the Henschel works at Kassel. It was hardly a new design, being an amalgam of features derived from many earlier trials vehicles and especially the VK36.01 chassis. This was now married to the 88mm turret Krupp had designed and manufactured for the losing Porsche submission in the heavy tank competition. The need to get the Tiger to the front as soon as possible precluded Henschel adopting mass production line methods as employed by the US for its M4 and the Soviets for the T-34. This in turn meant that the Tiger could never be available in the numbers required to become a mainstream tank in the panzer divisions, and would lead, as we shall see, to its employment in a different form of organisation.

The Tiger I was in production exclusively at the Kassel works between April 1942 and August 1944 and in that time 1,355 machines were manufactured, although many other companies in Germany were involved as subcontractors; major components of the tank such as the hull, turret, gun, optics and tracks were

The first series Tiger VI prototype.

Dr Erwin Aders (on right), head of Henschel's Panzer programme and the Tiger's chief designer, on a tour of Shop 5, 5 September 1942.

produced elsewhere. Just 83 Tiger Is were manufactured in 1942, 649 in 1943 and 623 in 1944. Each received a *Fahrgestall* or chassis number, with the first designated as 250001 and the last in August 1944 as 251346. The highest monthly production figure was 104 in April 1944. Compare this to the production of the M-26 Pershing, which was the US Army's attempt to match the Tiger I, with 1,436 being manufactured at two plants between November 1944 and June 1945.

The Tiger was a complex machine. Its manufacture was very labour intensive – each machine requiring 100,000 man hours to produce. It was also very costly relative to the Panzer Mark IV and the medium Panther. The former, which was the workhorse of the panzer divisions throughout the conflict, cost 103,463RM,

the latter 117,100RM, with the Tiger coming in at 250,800RM. Some measure of this cost can be gleaned by attempting to place this amount in today's prices.

In 1941 values, the Tiger would have cost $100,000 dollars. Allowing for inflation this

A Henschel workers use the horizontal six spindle borer to finish the Tiger's suspension holes.

⏶ The final drive holes are finished.

⏶⏵ Hulls which have reached Step 4 in the manufacturing process.

would equate to approximately $1,282,051 today. Although this is much less than the contemporary main battle tanks of the western powers, all of which are in excess of $4 million, it nonetheless represented a major financial investment for the German Army at the time. It is small wonder that great effort was expended in the field to recover damaged or broken down machines. The value the Tiger I represented as an extremely potent weapon was certainly matched by its cost! This did however also represent a very pragmatic approach by the Germans. Knowing that they could

◄ *Workers make alignments to ensure the turret ring is machined accurately by the vertical two-tool lathe.*

◀◀◀ *Fuel tanks, torsion bars and radiator fans wait to be installed on the finishing line.*

◀◀ *Tiger hulls enter the finishing line. Their empty engine compartments will soon hold Maybach motors.*

◀ *Overview of the finishing line.*

23

➤ *A worker welds on the front of the hull as suspension installation begins.*

➤➤ *A rear idler arm is installed using a hydraulic trolley.*

➤➤ *Installing the long torsion bars.*

◀◀ *A Maybach engine is brought in by an overhead crane.*

◀ *The engine is guided into the engine compartment.*

➤ *A roadwheel is guided into place.*

➤➤ *Final welding is performed on these early style roadwheels.*

➤ *The hull is readied for installation of the tracks.*

➤➤ *Workers use a steel cable wrapped around the drive sprocket to pull the track into place.*

never match enemy tank production it was believed that a superior machine like the Tiger would go far to offsetting enemy numerical superiority. At the time it made its appearance on the battlefield in Russia and Tunisia it was without doubt the most powerful and sophisticated tank in the world and represented a gun/armour combination that was not approached until the appearance of the Soviet JS-II on the Eastern Front in the early summer of 1944.

Nonetheless, the Tiger had been designed and envisaged as serving the role of a breakthrough tank in offensive actions. However, by the time it saw service the high tide of German conquests had passed and it was employed primarily in a defensive role. A Tiger battalion was rarely ever committed as a unit. It was far more common for companies to be hived off

◄ *An employee works in the driver's hatch as the tracks are fitted.*

▲ The biggest part
of TAKT 8 was the
installation of the turret.

➤ The turret basket
is clearly shown as the
turret is lowered onto
the hull by the overhead
crane.

➤➤ Workers watch as
the turret is almost in
place.

An almost complete Tiger, evidenced by the finishing touches such as the MG and muzzle brake covers, and the shovel on the front hull.

and distributed in penny packets or even individual tanks to provide fire support for infantry formations, as occurred in Normandy. Such was the prestige of the Tiger and its acknowledged prowess as a tank killer that they were always in demand

A Tiger close to completion. The stairs to its side were designed to give the assembly line workers easier access.

A completed Tiger is loaded onto a special railcar. The fender skirts were not installed at the factory because they would have made the Tiger too wide for transport.

when matters at the front became critical. This was certainly true in the East, where the deployment of a few Tigers could act as a breakwater in the face of large-scale Russian attacks, something borne out by the remarkable kill rates.

Any tank is always a compromise between three factors: firepower, protection and mobility. Choices have to be made as to which of these will dominate in the chosen design. In the case of the Tiger I, it was the first two which took priority. The combination of the firepower vested in the 88 mm KwK L/56 (KwK = *Kampfwagenkanone*) cannon married to the heavy armour of the machine rendered the Tiger by far and away the most formidable battle tank in the world between 1942 and 1944. It did however, have its limitations which inevitably had an impact on its employment.

Although the Tiger bore a superficial resemblance to the Panzers III and IV, it employed a novel method of construction. In contrast to earlier German tanks which employed bolted joints between the hull and superstructure, the hull of the Tiger was assembled using flat sections of heavy armour plate which were welded throughout. The panniers which had been added to the VK36.01 design extended out over the tracks and these accommodated the radiators for the Maybach engine. This was placed in a central section in the rear of the tank, and could be sealed for crossing deep rivers. Forward of this, the hull was divided into a further three compartments: the central fighting compartment and a forward section containing that of the driver on the left and on the right, that of the radio operator/machine gunner.

A 6 foot 1 inch turret ring carried the 88mm armed turret. The sides and rear of the turret was formed from a single piece of steel 82mm thick and bent into a horseshoe shape. Two 100mm thick bars were welded to the turret front plate with the gun in its mantlet, fitted between them.

Whilst elevation and depression of the main armament employed hand wheels used by the gunner, turret traverse was effected by a low geared, hydraulic system with a hand wheel backup. It took some 720 turns of the hand wheel to completely revolve the turret. The fighting compartment containing the commander, gunner and loader was suspended from the turret by three steel tubes and revolved

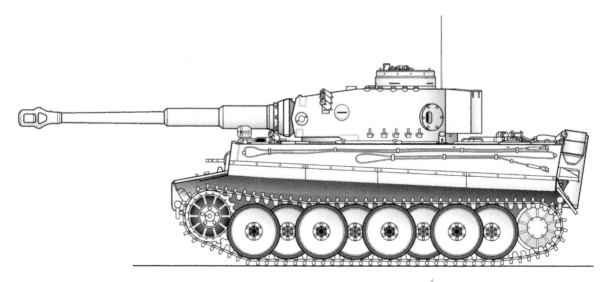

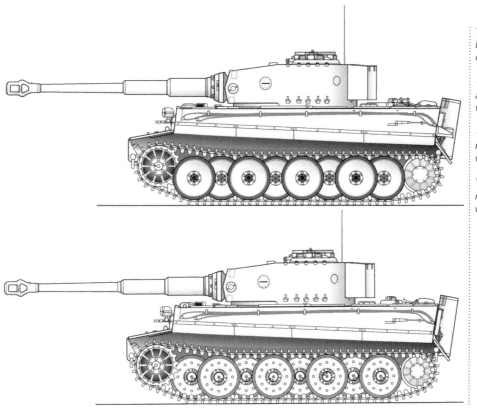

The main differences between the three models of the Tiger.

◄◄ The drum cupola and rubber rimmed road wheels.

◄ The cast cupola and rubber rimmed road wheels.

▼ The cast cupola and resilient steel road wheels.

33

34

with the turret. Until the latter part of 1943, the Tiger had a drum type cupola fitted with five vision slips, thereafter being equipped with the same lower profile, cast cupola fitted with six episcopes, as mounted on the Panther.

The Tiger I was also notable in being the first German tank to employ an overlapping and interleaved road wheel suspension. The triple road wheel layered arrangement initially employed steel discs with solid rubber tyres. These were later replaced with steel resilient road wheels that were internally sprung. Although the Tiger I, as with all German tanks, underwent an almost continual process of minor modifications on the production line in the light of combat reports, the three main models can be identified by their use of the cast cupola and steel road wheels.

A Tiger company rigged for rail transport. Note that the narrow transport tracks have been fitted, the mudguards removed and the outer layer of road wheels have been stored to the rear of the tank on the flatbed. This scene was a frequent one as Tiger units came to be seen more and more as the German Army's 'fire-brigades', especially on the Eastern Front.

ARMAMENT

The heart of the Tiger I was its armament. As we have seen, the 88mm KwK L/56 was derived from the 88 mm FlaK 18 anti-aircraft gun which had been in service with the German Army since 1933. With a barrel length of 56 calibres and a breech mechanism similar to that used in the L/43 and L/48 guns of the Panzer IV and assault guns, the Tiger I was the only vehicle to employ this weapon. All other German AFVs mounting the 88mm gun employed the Pak 43/1 KwK L/71.

Being a high velocity, flat trajectory weapon, the 88mm gun was extremely accurate with the ability to penetrate at battle ranges all Russian tanks it encountered with relative ease through to the appearance of the JS-II in the spring of 1944. The appearance of the latter with its massive 122mm gun permitted the Soviets to outrange the Tiger, being able to penetrate the frontal armour on the turret at over 1,000m.

On the Allied side, the American Sherman and the British Cromwell, and even the heavily armoured Churchill tank, all proved vulnerable to the 88mm gun, from Tunisia through to war's end, and it was for this reason it remained a real 'bogey' for Allied tankers. American and British tank crews learnt through bitter experience not to attempt to take on the Tiger frontally. It was a profoundly unnerving experience to close to 500m in a Sherman or Cromwell and still see AP shot bouncing off the frontal armour of the opponent. Allied tank crews learnt that it was best to stalk the Tiger, taking advantage of its slower turret rotation, the faster speed of their own machines and the deft use of cover to approach as close as possible and then fire at its more vulnerable sides. The equaliser for the Allies, albeit available in only small numbers on D-Day, was the modified M4A4 Sherman Firefly – its turret having been changed by the British to carry the extremely potent 17-pounder anti-tank gun. This weapon could penetrate the frontal armour of the Tiger's hull and turret at 1,700 yards and the rest of the turret and hull armour out to 2,500 yards. Small wonder that in Normandy and after, German tankers sought to take out Fireflies first.

The effectiveness of the penetrative power of the 88mm L/56 changes with the type of ammunition employed. The normal load carried by a Tiger was 92 rounds (although these were often increased by carrying extra rounds at the whim of the crew, especially when combat was intense and re-supply in the field was difficult). The ammunition with the greatest penetrative

Ammunition Designation	Ammunition Type
Panzergrenate 39 (Pzgr 39)	Armour piercing, ballistic capped with filler and tracer. Normally 50% of ammunition carried
Sprenggrenate 50 (Spgr 50)	High explosive shells. Normally 50% of ammunition load carried
Panzergrenate 40 (Pzgr 40)	Rarely used/carried. High velocity tungsten core
Grenate 39 HL (Gr 39 HL)	Occasionally carried. High Explosive Anti-tank (HEAT) employing a hollow charge warhead

An overall sand yellow Tiger I of s.Pz.Abt 502 undergoing training in France in June 1943 prior to being despatched to serve with Army Group North on the Eastern Front. The commander was Oberstleutnant Boelter who was to be credited with the destruction of 139-plus enemy tanks through to 1945.

power was Panzergrenate 40, which had a dense tungsten core (common in modern tank ammunition), but the metal was always in short supply in Germany and the ammunition was reserved for combating the most heavily armoured tanks and assault guns. From the summer of 1943, it was employed mainly against Russian heavy assault guns and the JS-II, the heavy frontal armour of which had been designed to cope with both the 88mm of the Tiger and the high velocity 75mm of the Panther. Panzergrenate 39 was adequate to deal with Allied tanks. The 88mm had good optics that permitted the Tiger to achieve first round hits beyond 1,000m. Naturally, the greater the range the more accuracy fell off, but it is instructive that the range scale on the sighting for the Pzgr 39 was graduated in units of 100m out to a maximum range of 3,000m and for the Spgr 50 out to 5,000m.

Until the advent of the Soviet JS-II and the British 17-pounder-equipped Sherman Firefly in 1944, it was the protection bestowed on the Tiger by its 100mm frontal armour and turret mantlet armour of 110mm that made it such a difficult target to destroy. The T-34 – which had proven so dangerous in the first two years of the Russian campaign, and which could penetrate the armour of all German tanks in service during that period – could not defeat the Tiger. Until early 1944, all T-34s were armed with a 76.2mm gun which was a variant of the standard army field gun that also doubled up as the Red Army's primary anti-tank gun. Fielding the Tiger with its potent gun/armour combination rendered the T-34 and KV-1 – which was encountered less often by the Germans after 1942 – highly vulnerable.

However, the Soviets were extremely capable of placing and camouflaging their 76mm anti-tank guns so that they could not be seen by Tigers until the range had closed sufficiently to negate the advantage of their armour. Michael Wittman, the much publicised Tiger ace, often said that he feared enemy anti-tank guns far more than he did enemy tanks.

Did you know?

Until August 1943, when the capability was deleted, the Tiger was equipped to undertake the fording of rivers up to 4.5m deep. For this they were equipped with special rubber seals, covers and plugs for various parts of the turret and hull. A four part telescoping tube was also provided that allowed air to get to the engine if the tank was underwater.

Armour: Hull	Thickness of Armour (mm)	Angle of Armour (°)
Nose	100	66
Vertical front plate	100	80
Lower sides	60	90
Upper sides	80	90
Vertical back plate	82	82
Top	26	
Bottom	26	

Armour: Turret	Thickness of Armour (mm)	Angle of Armour (°)
Mantlet	110	
Front	100	80
Sides	82	90
Back	80	90
Roof	26	

Tiger from the 1st Company s.Pz.Abt 503, Eastern Front, 1943.

Cutaway of a Tiger. (Courtesy Osprey Publishing, New Vanguard 5 – Tiger I Heavy Tank 1942–45, Thomas Jentz, Hilary Doyle and Peter Sarson)

This edited report by a lieutenant serving in s.Pz.Abt 503 in February 1943, gives a graphic impression of the damage that could be inflicted on a Tiger and yet for it still to remain operational. The battalion was serving in southern Russia at the time.

After two days of intense combat the crew of the Tiger

… counted 227 hits from anti-tank rounds, 14 hits from 57mm and 24 mm anti-tank guns and 11 hits from 76.2

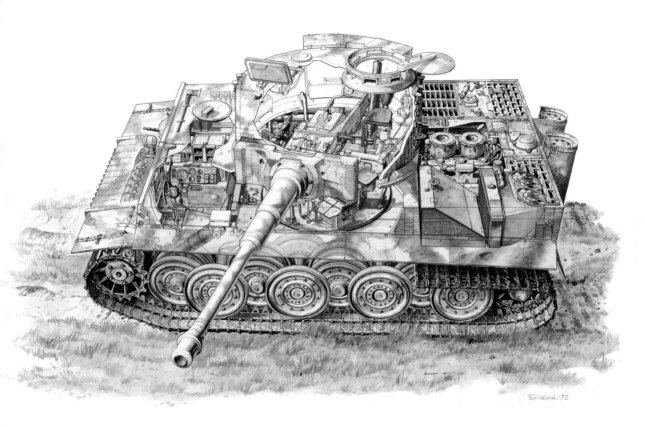

43

mm guns. This included two hits from 45mm anti-tank shells and 15 hits to the cupola. The right track and suspension were heavily damaged. Several road wheels and their suspension arms were perforated. The idler wheel had worked out of its mount. In spite of all of this damage, the Tiger still managed to cover an additional 60 kilometres under its own power … it can be said that the armour on the Tiger can withstand the most intense punishment the enemy can deliver. The crew can head into combat secure in the knowledge that they are surrounded by sufficient armour to keep out the most determined anti-tank round.

While there was no such thing as an invulnerable tank in the Second World War, it is apparent from this account why Tiger crews at least saw themselves as having a better chance of survival on the battlefield than those serving in other types.

In the hands of an experienced driver the Tiger I was a fairly reliable machine. But, much as with the Panther medium tank, the exigencies of war had forced the commitment of this new heavy panzer into combat before many of its technical bugs had been worked out.

The most common bug arose from problems with the transmission and final drives stemming from their having been originally designed for the VK36.01 medium tank. Coping with servicing a tank a full 20 tons heavier when in combat frequently led to overstress of these components and their subsequent failure. This same reason accounts for problems arising with the gears and clutches, which could be stripped when handled roughly. Long road marches were not suited to the Tiger but in the absence of rail transport the need to have them available as a 'fire brigade' often saw them having to carry these out. On 5 August 1944, s.Pz.Abt 501 operating on the Eastern Front, reported that most Tigers in two of the companies participating in a 50km road march had broken down due to final-drive failure. Although some of these problems were never completely eliminated, feedback from the front to Henschel saw modifications introduced on the production line.

One of the reasons why Tigers were grouped in their own battalions was the realisation that such a complex panzer required specialised equipment that would be far more economically employed servicing a larger number of machines. For example, failure of the aforementioned transmission and final drives required a convoluted process of repair which

> *Tiger crews were well aware that to ensure the reliability of their charges they needed to make frequent halts to check the engine and final drive. The crew of this Tiger of s.Pz.Abt 503 have just raised the hatch covering the Maybach engine to check all is well, Summer 1943.*

necessitated the complete removal of the turret. This is in itself was a heavy object and the *Werkstatt-Kompanie* of Tiger battalions were issued with 15-ton Fries cranes that had to be assembled in the field, behind the Front

Servicing the engine was also problematic. Its very compact fit made access to the Maybach HL 210 engine – as first fitted it developed 650hp with an improved 700hp HL230 P45 used from the end of 1943 – difficult and led to fires as cooling air had difficulty circulating in the narrow compartment. On occasions, such fires led to the total loss of the Tiger. Although modifications reduced the frequency of such problems, they could on occasion lead to disastrous outcomes. On 3 January 1944 s.Pz.Abt 503, which had been totally worn down by the winter fighting, received 45 new Tigers as replacements. Of these, the engines of 11 caught fire, with one tank being destroyed. The report also noted the tanks had many 'other deficiencies'. The latter was presumed to have been caused by sabotage with steel shavings having

◄ *Mechanics prepare to the lift the turret from an early model Tiger I using a Fries crane.*

➤ Two 18-ton FAMO half tracks tow a broken down Tiger.

been found in the idler wheel mounts – a consequence of the employment of slave labour in factories servicing different parts of the production process.

Although other factors – such as a shortage of spare parts and fuel – had an impact on serviceability, the biggest problem for the Tiger was recovery from the battlefield. When fully loaded the combat weight could vary between 56 and 58 tons. Damage to the tracks from mines was a frequent occurrence. However, there was no vehicle in service with the *Heer* that could recover a broken down Tiger. Two, sometimes three Famo 18-ton half-tracks were used to pull a Tiger employing a rigid towing bar. Later, some battalions were issued with the Berge Panther – a recovery tank based on the medium Panther chassis. Although Tiger crews were forbidden to recover another damaged Tiger with their own, this order was often ignored.

Many Tigers were in consequence perforce abandoned or destroyed by their crews. This was also prompted by a lack of fuel. 25 May 1944 was a disaster for s.Pz.Abt 507 which was serving in Italy. Seven Tigers had to be blown up for this reason. A further 12 were lost the same day for this or other reasons. The battalion commander was dismissed.

It could be inferred that the Tiger was a dubious proposition; that would be wrong. Tanks are always complicated, sophisticated and difficult to maintain. Some are just more so than others. On the one hand, the US M4 was vulnerable to the fire of most German AFVs in 1943–45, but automotively, most variants of the type were a far easier proposition to work with than the Tiger. In the end, Tiger repair crews, in all probability the unsung heroes of this story, performed prodigious feats to keep their charges in the field. It was far from an easy task, but they knew that when running and engaging the enemy, their charges were capable of inflicting massive losses.

There were two variants of the Tiger I. From March 1943, the most common was designated PzBefWg Tiger. This was the *Befehlswagen* – a command variant of the heavy panzer – of which 84 were produced. These were modified internally to carry more radio transceivers, the extra space being generated by reducing the

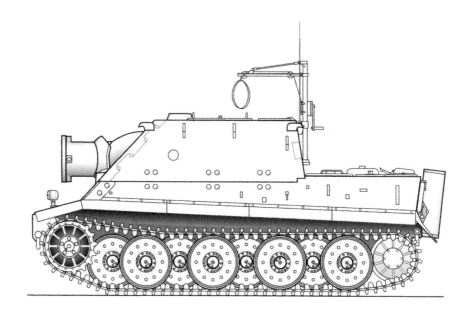

➤ *A Sturmtiger, designed for combat in urban areas.*

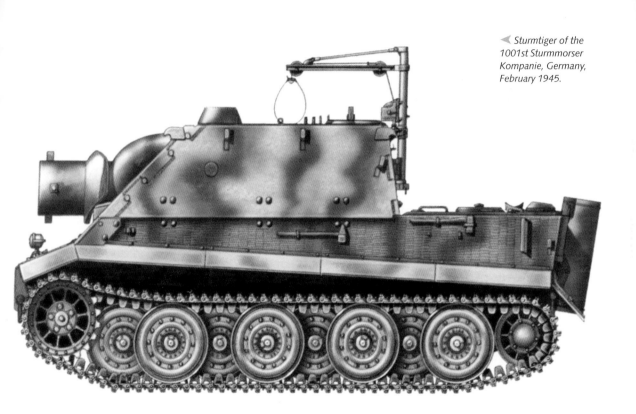

Sturmtiger of the 1001st Sturmmorser Kompanie, Germany, February 1945.

main ammunition storage to 66 rounds and by deleting the co-axial machine gun. The most common way of differentiating the two models of this variant from the standard Tiger was that they carried more radio aerials, one being on the roof and the more prominent being a star antenna mounted on the engine deck. The radio fit was different in each model.

The other variant on the Tiger I chassis was the Sturmtiger, of which only ten were produced in 1944. The vehicle was a response to the need for a heavily armed assault tank that could employed in fighting in built-up areas of the type experienced by the German Army in Stalingrad. Although originally intended to mount a 210mm howitzer, the lack of such a weapon led to the adoption of the Raketenwerfer 61 L/54. This was a 380mm weapon employing a rocket projectile with each shell weighing 344kg. The Tiger 1E chassis was selected and the firm of Alkett in Berlin was contracted to produce a small number. The rocket projector was mounted in fixed superstructure and when fully loaded the machine weight 70 tons and had a crew of seven. The size of each shell meant that only 12 could be carried. Although there exists film showing one being used in the fighting for Warsaw in 1944, the type made no impact at all on the battlefield although a few were encountered in the advance into western Germany by the Allies in 1945.

The Tiger I was an interim, albeit very successful design. Ever since the meeting with Hitler in May 1941, it had been the intention to arm a heavy panzer with a version of the Flak 41. As we have seen, the Krupp turret originally designed for the failed Porsche submission for the Tiger I and then fitted to the production Henschel machine, was too small to mount the larger weapon. To do so would require a new design and with the Tiger I having been in production for just a few months the *Heereswaffenamt* (Army Ordnance Department) issued a specification for a new heavy tank mounting the 88mm L/71 gun with heavier and sloped armour as was being employed on the new Panther medium tank. Once again, Porsche and Henschel went head to head. As with the initial Tiger design, the Porsche submission known by the designation VK45.02(P), continued to employ the problematic petrol-electric engines, with one powering each track. Porsche had taken his original design and modified it for the new specification by sloping the armour and redesigning the chassis to take a larger turret for the 88mm L/71. Once again Henschel won out, with the Tiger II going in production at Kassel in December 1943, alongside the Tiger I. The first production Tiger II left the Kassel factory the following month.

Although Porsche's heavy tank was cancelled in November 1942, Krupp was already in the process of building a turret for his failed design and it was decided that as materials had been already collected to undertake limited production, 50 of the P *turm* – the 'Porsche' turret – would be built and mounted on the Henschel VK45.03 chassis. Thereafter, every Tiger II, from the 51st onward, would have the armament mounted in a new Henschel-designed *Serienturm* or series turret. This was not only easier to manufacture than the P-turm, but also provided far better protection. The P-turm had a shot trap under the curved front gun plate which allowed an enemy shell to be deflected down into the hull armour where the radio operator and driver were seated. The new series turret decreased the frontal area, eliminating the shot trap and incorporated a more ballistically sound bell mantlet that housed the 88mm barrel.

▼ *The first 50 Tiger IIs were built with the P-Turm.*

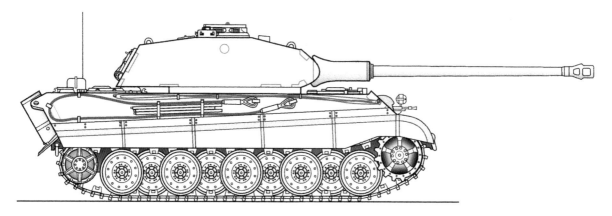

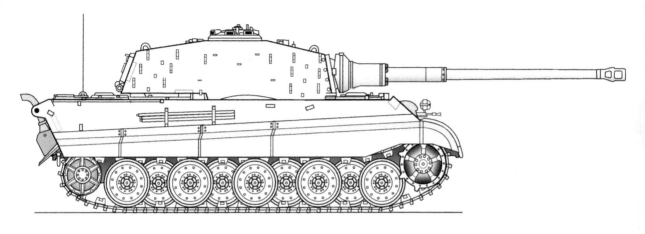

An abortive attempt in early 1943 to ensure that the Tiger II and the Panther II – the projected successor to the first generation Panther – shared common parts came to nought when the latter was cancelled. It did however lead to a delay in getting the Tiger II into production. Until 16 March 1943, the new heavy tank went by the designation of Tiger H3. From that date it was known as the Tiger II. The oft employed name of *Koenigistiger* or King Tiger was never an official title. In official parlance, the new heavy panzer was designated Panzerkampfwagen VI Ausf B (Sdkfz 182).

△ Later model Tigers were fitted with the Henschel-designed Serienturn.

Between December 1943 – when the pilot vehicle was produced – and March 1945 when the very last Tiger II left the production line, mere days before the Americans captured the Kassel plant, 484 of the heavy panzers had been manufactured. This was however, far fewer than had been planned. In February 1944, just 5 Tiger IIs were produced, whereas 95 Tiger Is left the production line. By September, when production of the Tiger I had ceased, output of the Tiger II had climbed to 94 even though 100 were planned for. The anticipated output of 145 per month scheduled for December was set back when the Henschel works was hit by a series of heavy bombing raids in late September and early October. This resulted in the destruction of 95 per cent of the plant. In consequence, just 26 Tiger IIs were completed in October and the same number in November. The factory had recovered sufficiently to produce 56 in December but thereafter, as supplies from the main contractors and subcontractors slowly broke down, only another 112 were produced before the end of March. In the seven months between September 1944 and March 1945, only 283 out of the planned 940 Tiger IIs left the production line.

FIREPOWER

The 88mm KwK 43 L/71, which was the most powerful gun mounted on a German tank in the Second World War, was not a weapon unique to the Tiger II. It was the main armament of the Ferdinand/Elefant self propelled gun and the Hornisse/Nashorn tank destroyer. It was also the main armament selected for the Jagdpanzer V Jagdpanther. The performance of this weapon endowed all of these vehicles with outstanding tank killing capability.

The KwK 43 was derived from the PaK 43, which was the best German anti-tank gun of the war, albeit very large, and when mounted on the carriage of the 105mm howitzer as an expedient, extremely heavy and cumbersome – traits that earned it the nickname *Scheunetor* or 'barn door'. However, a measure of its effectiveness is found in a report that credited one gun with destroying six T-34s at a range of 3,500m. Nor did it require the special tungsten-cored ammunition to effect a kill of the more heavily armoured soviet tanks. Another account describing the PaK 43 in action on the Eastern Front reported that

… penetrative performance of the Panzergrenate 39 is satisfactory at all ranges, so that all enemy tanks appearing in this sector – the T-34, KV-1, JS-II – could be engaged with destroying effect. On being hit, the tanks showed darting flame three metres high and were burned out. Turrets were mostly knocked off or torn away. A T-34 was hit from the rear at a range of 400 metres and the engine block was flung out a distance of five metres and the turret cupola for 15 metres.

The KwK 43 of the Tiger II boasted this same performance. With a barrel length of 71 calibres – approximately 21 feet long – it was the largest calibre weapon employed on a panzer. Early models carried a one-piece barrel later replaced by one constructed in two parts in May 1944.

Although the cartridge cases for the KwK were modified, the ammunition types were the same as those fired by the PaK43. Of the 78 shells carried by the Tiger II, 22 were stored in a bulge in the turret rear with the nose of the projectile facing forward to permit ease of loading. A further 48 rounds were mounted in horizontal panniers on either side of the hull. Two machine guns were mounted, one operated by the radio operator via a ball mount in the glacis, and the other in the turret. A third could also be carried on a special attachment fitted to the cupola ring and used by the tank commander for anti-aircraft defence.

The effectiveness of the firepower of the Tiger II was often negated by its slow turret traverse. The time taken to turn through 360 degrees, even when powered, could be up to 75 seconds. In the event of this needing to be done manually, the loader required 680 turns and the gunner 700 turns to effect a full turn. In that time, more nimble Russian and Allied tanks were able to stalk the vehicle by virtue of their higher speed and defeat the Tiger by flanking shots to the thinner turret and side armour.

ARMOUR

The armour on the Tiger II was better arranged and was also much thicker across its frontal arc than on the Tiger I. While it had been intended that it would have sloped armour as per the Panther medium tank from the outset, the design had been delayed to accommodate Hitler's insistence in January 1943 that armour be strengthened, with that on the hull front being raised to 150mm and the sides to 80mm.

The frontal armour of the Tiger II made it virtually invulnerable to the fire of any

Armour for the Serienturm Tiger II (Henschel)		
Armour Location on Hull	Thickness of Armour (mm)	Angle of Armour (°)
Front glacis plate	150	50
Front nose plate	100	50
Superstructure sides	80	25
Hull side plates	80	0 vertical
Tail plate	80	30
Deck plates	40	90 horizontal
Front belly plate	40	0 horizontal
Rear belly plate	25	0 horizontal
Turret – front plate	180	
Sidewalls	80	
Roof	40	

enemy tank. Even the JS-II was unable to penetrate the frontal hull armour at 2,600m whereas at the same distance, the Tiger II could defeat the Soviet tank. When Tiger IIs were successfully defeated by Russian and Allied tanks they had been 'stalked' and killed at close range by a penetration of the thinner turret or hull sides.

➤ *A still from a propaganda sequence from the weekly newsreel 'Deutsche Wochenschau' taken in December 1944 shows Tiger IIs of s.Pz. Abt 503 drawn up for a review which is being taken by Lieutenant von Rosen.*

MOBILITY

The Achilles heel of the Tiger II was its poor mobility due to its being underpowered. It employed the same Maybach HL 230 P45 engine as used on the Tiger I, even though a combat-laden Tiger II was 11 tons heavier. It also suffered from engine fires and the same problem with its drive train as the Tiger I. While breakdowns were frequent, it would be wrong to infer that the Tiger II was nothing but a very slow armoured fortress. When operating efficiently, it was a formidable fighting machine with acceptable mobility, being able to sustain a cross-country speed of between 15–20 kmph. But speed by the second half of 1944 – when it entered service – was of less concern to a German Army on the defensive. The combat effectiveness of the Tiger II was vested in its gun which was best employed in taking on the enemy at great distance. Fast speed in such circumstances was not necessary whereas the capability to reliably move from firing point to firing point and into cover was.

As with the Tiger I, a number of Tiger IIs were outfitted as command panzers. The extra transceivers were placed in the rear of the turret thus reducing the number of 88mm rounds carried.

THE *JAGDTIGER*

As with the Panther, the *Heereswaffenamt* decided that a tank destroyer version of the Tiger II would be produced. Although designated the *Jagdpanzer VI, PanzerJaeger Tiger Ausf B*, it was better known as the *Jagdtiger*. It had the twin distinction of mounting the most powerful anti-tank gun as well as being the heaviest AFV to enter service in the Second World War.

The heart of this particular tank hunter was the massive 128mm PaK44 L/55 anti-tank gun. It was able to penetrate 230mm of armour at 1,000m and 173mm at 3,000m, and could defeat any Allied tank. The weapon itself was carried by a slightly lengthened Tiger II chassis and was housed in a fixed superstructure, the frontal armour being 250mm thick and the sides and rear 80mm thick.

When fully loaded, the Jagdtiger weighed 70 tons, in consequence of which

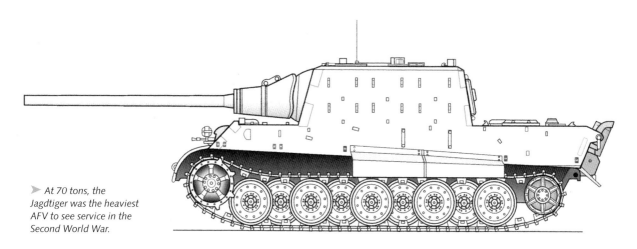

➤ *At 70 tons, the Jagdtiger was the heaviest AFV to see service in the Second World War.*

it suffered from the same mechanical problems as the Tiger II. Unsurprisingly, it suffered from many breakdowns. Although orders were placed for 150, only 48 were built and these equipped two battalions. These were encountered by Allied troops from December 1944 onwards in Belgium and Germany. There is a famous short film sequence of the Jagdtigers of the Jg.TgrAbt 512 surrendering in the town of Iserlohn to the Americans in April 1945. It shows just what an impressive machine the Jagdtiger was. For those who wish to see a Jagdtiger, the Tank Museum at Bovington has one on static display. It was the only one fitted with the Porsche suspension.

◄ A fine view of a Jagdtiger which clearly shows the massively armoured structure housing the 128mm anti-tank gun atop the chassis.

We have already seen that the low production output of the Tiger meant that it could not be employed to re-equip the panzer divisions. It was therefore decided that the tank would be allocated and employed by newly created independent tank battalions. These were designated by the abbreviation s.Pz.Abt – meaning *schwere Panzer Abteilung* – or heavy tank battalion. Between May 1942, when the first battalion was formed, and 1945, the army formed ten independent tank battalions. These saw service in Russia, Tunisia, Italy and Western Europe.

In 1942, and into early 1943, the shortage of Tigers saw the first five battalions – 501 through 505 – partially equipped with Panzer IIIs to augment their numbers. The need to retain Tigers for training purposes and satisfy the competing claims made on the few machines leaving the production lines saw their numbers spread thinly between the army and the Waffen SS. Between October 1942 and March 1943 these five formations were equipped and committed to action in the mixed Tiger/ Panzer Mark III organisation.

s.Pz.Abt 501	20 Tigers	24 Panzer IIIs	October 1942
s.Pz.Abt 502	19 Tigers	27 Panzer IIIs	December 1942
s.Pz.Abt 503	22 Tigers	29 Panzer IIIs	December 1942
s.Pz.Abt 504	20 Tigers	25 Panzer IIIs	February 1943
s.Pz.Abt 505	20 Tigers	25 Panzer IIIs	March 1943

Heavy tank battalions 506 through to 510 only ever employed Tigers. It was only in early March 1943 that orders were issued to expand each heavy tank battalion to 45 Tigers, and it was this structure that was maintained through to the end of the war whether or not by late 1944 or 1945 the respective battalion retained the Tiger I or had been re-equipped with the Tiger II.

It was never possible, however, to ensure that this number was always available. Combat losses, difficulties with supply and bombing raids on factories saw numbers fluctuate throughout the period.

Heavy Tank Battalions		
(*schwere panzer abteilungen*)	Formed	Disbanded
s.Pz.Abt 501	May 1942	February 1945
s.Pz.Abt 502	May 1942	May 1945
s.Pz.Abt 503	May 1942	January 1945
s.Pz.Abt 504	January 1943	May 1945
s.Pz.Abt 505	February 1943	April/May 1945
s.Pz.Abt 506	May 1943	April 1945
s.Pz.Abt 507	September 1943	May 1945
s.Pz.Abt 508	September 1943	February 1945
s.Pz.Abt 509	September 1943	May 1945
s.Pz.Abt 510	June 1944	May 1945

The only *Heer* formation to employ an organic Tiger unit as part of its organisation was the elite Panzer Grenadier Division *Gross Deutschland* (hereafter GD). Initially allocated the designation of 13th Company of the Panzer Regiment, this formation served in this fashion until 1 July 1943, when it was re-designated the 9th Company. Two new Tiger companies, namely the 10th and 11th were added to give GD is it own organic Tiger s.Pz.Abt which was then designated the 3rd battalion of the Panzer Regiment. This formation served solely on the Eastern Front through

▼ *Late model Tiger IE of GD destroyed in the fighting for Konigsberg, February 1945.*

to the destruction of its last Tiger I on 19 March 1945. GD never employed the Tiger II.

In addition a number of other Tiger units were raised either as extemporised formations or they served in specialised roles. In many cases they only deployed one or a few Tiger Is. In the latter category two Tiger units equipped with the Tiger I and later the Tiger II in the role of radio-control command vehicles (*Funklenk* = Fkl) for B IV demolition vehicles. They were designated *Panzerabteilungen* (Fkl) 301 and 316 (Fkl).

PzKpfw VI Kfz 182 Tiger Ausf B from the 316th Panzer Kompanie (Fkl), attached to the Panzer Lehr Division, Normandy, 1944.

WAFFEN SS TIGER UNITS

It was decided by Hitler early on that the Waffen SS would receive their own organic Tiger companies. In the summer of 1942, all three divisions – the Leibstandarte Adolf Hitler, Das Reich and Totenkopf – had been withdrawn from the Eastern Front and sent to France for rebuilding where they were upgraded to panzer grenadier divisions, although they were in practice – given the numbers of tanks with which they were equipped – *de facto* panzer divisions. The Tigers equipped the heavy tank companies of SS Panzer Regiments 1, 2 and 3.

SS.Pz.Abt 101 (501)	July 1943	May 1945
SS.Pz.Abt 102 (502)	April 1943	May 1945
SS.Pz.Abt 103 (503)	July 1943	May 1945

In July 1943, it was decided to raise an SS heavy tank battalion for service with the newly created 1 SS Panzer Corps. It was to have 45 Tiger Is as per the army battalions. Two others were also raised. The first two of these saw extensive combat through to war's end and extensive service on the Eastern Front and both were engaged in Normandy.

On 10 November 1942, following the Allied landings in North Africa, Hitler ordered that s.Pz.Abt 501 be shipped to that theatre. The Tigers of 1/501 were transported by rail to Sicily, then by ferry to Tunisia, with the first three offloading at Bizerte on 23 November. The second battalion had been delayed as it had been employed in the German occupation of Vichy France.

In the meantime, the first Tigers and supporting Panzer IIIs were employed as a

Did you know?

While not the only surviving Tiger I, the example held by the RAC Tank Museum at Bovington, Dorset, England is the only running Tiger in existence. Issued the Fahgerstall number 250112, it was the 122nd Tiger IE built by Henschel at its factory in Kassel and left the production line in February 1943.

It was issued to s.Pz.Abt 504, this unit having been established in Fallingbostel (post-war a major base for the BAOR) on 18 January 1943. On 17 March, the battalion was ordered to make ready for deployment in Africa. It took until 16 April for the last Tiger of the battalion to be offloaded in Tunisia. Tiger 131 was knocked out by British Churchills of No:4 Troop, A Squadron of 48th Royal Tank Regiment at Medjez el Bab. It was recovered and shipped back to the UK for testing. In 1951, it was handed over to the Tank Museum and in 1992 the decision was taken to return it to working order, which took many years of hard and expensive work.

A Tiger IE knocked out by the British Army in Tunisia, March 1943.

kampfgruppe (battlegroup) alongside 10th Panzer Division and were involved in the Battle of Tebourba against US armoured units, which lost 134 tanks. By mid-January 1943, 501 was in place with its two companies at full strength with 20 Tigers and supporting Mark IIIs. However, their subsequent deployment was to presage the manner in which Tiger units would be fielded for much of the rest of the war. Rather than committed as a battalion, the two companies were split up and allocated to support other formations. Even in the short time the Tiger had been operating in Tunisia, its combat value had come to be greatly valued as a battle winner with ample evidence that its appearance had an impact on the enemy both materially and psychologically. s.Pz.Abt was subsequently involved in the Battle of Kasserine. On 17 March the eleven surviving Tigers were attached to the s.Pz.Abt 504.

Only the headquarters detachment and the first company of s.Pz.Abt 504 arrived in Tunisia with its mixed Tiger and Pz MkIII complement over January and February 1943. The second company was retained in Sicily where it eventually fought following the Allied invasion of the island on 10 July. This unit was involved in very heavy fighting through to the German surrender, taking a toll of Allied tanks with some 90 being claimed on 6 May. The remnants of both s.Pz.Abt 501 and 504 surrendered on 12 May. Both battalions were rebuilt in Germany.

Did you know?
From November 1942 Tigers intended for service in North Africa were equipped with an air filtration system called *Feifel*. This was designed to stop fine sand entering the engine. You can identify such Tigers by the air filters which were mounted on the top corners of the hull rear. The system was discontinued in August 1943.

With s.Pz.Abt 502 committed to the fighting around Leningrad, the next unit sent to the Russian Front was s.Pz. Abt 503. This was despatched to southern Russia where it was involved in very heavy operations following the encirclement of Sixth Army in Stalingrad and helping to fight the Soviet winter offensive in the region. At the end of March the battalion had 31 Tigers on strength and on 10 May it was redeployed to Kharkov. In common with the other Tiger battalions so equipped, it now lost its Panzer IIIs and received the full complement of 45 Tigers. At the end of May it was subordinated to Army Detachment Kempf in preparation for the German summer offensive.

February 1943 had seen the arrival of the three divisions that formed the SS Panzer Corps in southern Russia. These

➤ *Tiger IE of the 13 Heavy Tank Company Grossdeutschland Division, March 1943.*

deployed their three Tiger companies. The nine Tigers forming the heavy company of GD also arrived in Russia from Germany at this time. All four companies were involved in the counter offensive launched by Field Marshal von Manstein, which had had recaptured Kharkov by the end of March.

The impact of the Tiger in this operation was clear to see. On 15–16 March the Tigers of GD destroyed 67 Soviet tanks. General Erhard Raus later described the impact of the Tiger:

It was in this action that PzKw VI Tigers engaged the Russian T-34s for the first time, and the results were most gratifying to us. For example, two Tigers acting as a panzer spearhead destroyed an entire pack of T-34s. Normally, the Russian tanks would stand in ambush at the hitherto safe distance of 1,200 metres and wait for the German tanks to expose themselves upon exiting a village. They would then take the tanks under fire while our Panzer PzKw IVs were outranged. Until now, this tactic had been foolproof. This time, however, the Russians miscalculated. Instead of leaving the village, our Tigers took up well-camouflaged positions and made full use of the longer range of their 88mm guns. Within a short time, they knocked out sixteen T-34s that were sitting in open ground and, when the others turned about, the Tigers pursued the fleeing Russians and destroyed more tanks. Our 88mm armour piercing had such a terrific impact that they ripped off the turrets of many T-34s and hurled them several yards. The German soldiers, witnessing these events immediately coined the phrase: 'the T-34 tips its hat whenever it meets a Tiger'. The performance of the new Tigers resulted in a great morale boost.

After the recapture of Kharkov and the close down of the German counter offensive towards the end of March 1943, both sides prepared for the coming summer operations.

A Tiger IE of s.Pz.Abt 502 serving with Army Group North in Russia, summer 1943.

The operation which has become synonymous with the employment of the Tiger I was the last great German offensive on the Eastern Front launched in July 1943. Given the codename Operation Citadel (*Unternehmen Zitadelle*), it was an abortive attempt to encircle, destroy and capture the huge Red Army forces deployed in the Kursk salient in the Eastern Ukraine.

While Citadel was to mark the largest assemblage of these formidable panzers ever employed in a single operation, only two Tiger battalions actually served in the offensive. Heavy Tank Battalion 505 was allocated to General Model's Ninth Army which was tasked with assaulting the northern neck of the salient. The remaining Tigers, including those of the second Army Battalion was allocated to Army Group South. This was the 503rd, which formed part of Army Detachment Kempf. This formation was at full strength on 5 July with 45 Tigers on hand, albeit with only 39 being committed to battle on that date, with the remaining 6 under repair. However, immediately prior to the offensive, the decision was taken to allocate one company to support the operations of each of the three panzer divisions of III Panzer Corps – the primary assault formation of Army Detachment Kempf. This was contrary to the inspector of tank troops' specific instruction to strive to employ a Tiger battalion as a coherent unit. Only in such fashion did he believe it possible to maximise the combat potential of the heavy panzer. It was only some days into the offensive that the 503rd was reunited and fielded as a battalion.

The remaining Tiger units were serving with Fourth Panzer Army on the southern neck of the salient under the command of Panzer General Hermann Hoth. These comprised the three heavy tank companies of the IInd SS Panzer Corps and the 13th Company of the GD Panzer Grenadier Division. In the former, the 1st SS.Pz.Gr.Div Leibstandarte Adolf Hitler was fielding 13 Tigers, the 2nd SS Pz.Gr.Div Das Reich was fielding 14 and the 3rd SS.Pz.Gr.Div Totenkopf was fielding 15. The 13th Company of GD was fielding 14 Tigers.

In total, 146 Tigers saw action during Citadel. While this figure constituted just five per cent of the total number of tanks and assault guns committed by the Germans to the offensive, it nonetheless amounted to some 60 per cent of the total stock of these machines available to the Heer at the beginning of July 1943. Nonetheless, the Tiger's reputation and the prestige that went with claiming to have destroyed one prompted the Red Army to state that many hundreds were destroyed during the course of the two-week offensive. These Soviet claims, however, exceeded by a considerable margin the total number of Tigers built to that date! Such was the potency of the Tiger in Soviet eyes, it became the primary symbol of the immensity of the German offensive effort in this offensive.

NINTH ARMY – THE NORTH OF THE SALIENT

On the opening day of the offensive the Tigers of the two companies of s.Pz.Abt 505 had destroyed 42 T-34s and had effected a breakthrough of the Soviet

15th Infantry Division. Two days later, the ferocity of the Soviet defences accounted for the total loss of three Tigers. By the 8th, just three Tigers were operational with the rest, minus those lost, in the hands of the repair teams. Two days later, with the third battalion having arrived from Germany, the battalion reported 26 available for service.

Seen from the 'other side of the hill', a Soviet encounter with a Tiger was later described by Ivan Sagun, who commanded a T-34 on the northern sector. His account detailed his 'duel' with a Tiger of s.Pz.Abt 505 on 7 July. He observed how the heavy panzer

… fired at me from literally one kilometre away. His first shot blew a hole in the side of my tank. With the second, he hit my axle. At a range of half a kilometre, I fired

at him with a special calibre shell, but it bounced off him like a candle. I mean, it did not penetrate his armour. At literally 300 metres, I fired my second shell. Same result. The he started looking for me, turning his turret to see where I was.

Determining on this occasion that discretion was the better part of valour, Sagun ordered his driver to reverse the tank and seek cover behind a screen of trees.

Tiger I of s.Pz.Abt 505 serving with Ninth Army during Operation Zitadelle.

The commander of the Central Front – General Konstantin Rokossovsky – later wrote that it was the losses being inflicted on his T-34s by the Tiger in the open that prompted him to have many of them dug-in and used as firing points.

Tiger 132 of s.Pz.Abt 505 in action at Kursk, 6 July 1943.

The fighting on 10 July reduced the number of Tigers on the following two days to 11 each, with 20 Tigers available on the 14th, and 16 two days after that. By that date all German offensive action had ceased as panzer units were pulled out of the salient to contest the massive Soviet offensive launched against the rear of Ninth Army in the Orel bulge; s.Pz.Abt 505 was redeployed in its turn on 19 July.

Also committed to the offensive, and serving only in the north of the salient, was s.JgPz.Abt 656 which was fielding the Ferdinand tank destroyer. Hitler expected much of this machine and while the two companies of Ferdinands did indeed destroy a sizeable number of Soviet tanks with their 88mm PaK 43 anti-tank guns, their employment was nevertheless attended by many problems. Once in combat, it

was found that the lack of an on-board machine gun made it vulnerable to Russian soldiers with magnetic mines. It had to fall back on using its 88mm gun for close protection. Although no more vulnerable to mines that the Tiger, the battle weight of the Ferdinand, approaching 70 tons, made recovery of a damaged machine extremely difficult. Indeed, a number had to be abandoned for that reason, with

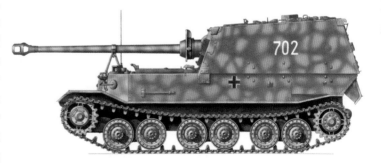

Panzerjaeger Sd Kfz 184 'Ferdinand' of the 7th company, s.Pj.Abt 654, Ninth Army, July 1943.

the Soviets making great propaganda play with them after the battle when they were extensively photographed.

The survivors were later withdrawn to Austria for rebuilding. These emerged with a commander's cupola and a built-in machine gun and renamed the 'Elefant'. These served in Italy and on the Eastern Front, with the very last being destroyed in early 1945.

➤ *On 8 July Soviet forces attacking the village of Alexandrovka captured the crew of Ferdinand 333 of the 3rd Company abt.653.*

➤➤ *Russian soldiers were photographed in front of their prize, and the image was published in* Pravda *and* Izvestia.

△ PanzerJaeger 'Elefant' of 614 s,Pj.Kp, Poland , January 1945.

4TH PANZER ARMY AND ARMY DETACHMENT KEMPF IN THE SOUTH OF THE SALIENT

Operating on the left of the 2nd SS Panzer Corps was the Grossdeutschland Panzer Grenadier Division, with its company of 13 Tiger tanks. GD saw very heavy fighting from the opening day of the offensive. By day's end on 5 July, just three Tigers were still operational although none had been lost. By the end of the 7th, all the Tigers in the company had fallen out and were under repair. Eight were returned to service on the following day, and were back down to three one day after that. The same see-saw in numbers of serviceable Tigers continued through to 16 July, when GD was pulled out of the line and sent north to support Ninth Army. On that day just six of the heavy panzers were still operational.

IIND SS PANZER CORPS

Under the command of Obergruppenführer Paul Hausser, the three divisions of the 2nd SS Panzer Corps were fielding three companies of Tiger Is with 42 on strength and 35 actually serviceable on 5 July.

The three companies were very heavily committed in the opening days of Citadel as the cutting edge of the German drive to force through successive lines of Soviet defences. The LAH lost one Tiger in the first two days, but accounted for 50 T-34s, one KV-1, one KV-2 and 43 anti-tank guns.

The experience of the other two SS heavy tank companies was similar. Das Reich had two Tigers knocked out by mines on the opening day but accounted for 35 T-34s and US M3s in the first two days of fighting. Totenkopf also had five Tigers immobilised on the 5th due to mines and was involved in extremely heavy fighting on the eastern flank of the Corps.

It is 1100 … This is the hour of the tank. Unnoticed we had assembled, the Tigers flanked by medium and light companies [Panzer IIIs, IVs and half tracks]. Our field glasses searched the horizon, groping in the smoke that covers the enemy bunkers like a curtain. The leader of our Tiger, an Obersturmfuehrer from the Rhineland

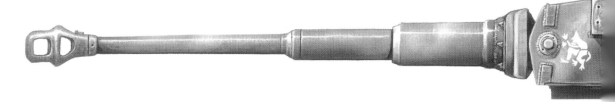

whose calmness ennobles us, gives the order to attack. The tank engines begin to howl [a distinctive feature of the Maybach engine] as we load the guns. The heavy tanks slowly roll into the battle zone. At 200 metres, the first anti-tank fires at us. With a single round we blow it up. All was quiet for a while as we rolled over the abandoned enemy trenches. We waved to our brave infantrymen from our open hatches as we passed them …

Enemy infantrymen ran through the corn trying to get away from us … Our machine gunners fired at them and forced them to take cover. A heavy enemy truck was seen in the woods to our right attempting to escape. We fired upon it and it burst into flames …

It was now 1200. The sun was burning hot, so we have opened the hatches and are peering at the terrain ahead. We do not receive effective fire until an hour later when we see two T-34s stationed on the dominating heights to the north. Their first shells land near us. Then a number hit our frontal plate and hull, but they do not do much harm. We move forward a little to take up a battery firing

Turrent profile of a Tiger IE of the 2nd SS Pz.Gren.Div Das Reich. The gnome insignia was carried by the Tigers of the 8th Company of the heavy panzer regiment. The turret number is idiosyncratic. The 'S' stood for schwere – 'heavy'. The two-digit number identifies it as the first machine of the second platoon.

position. Load, unlock, shoot! A hit! We pursue. The first T-34 is burning. Our neighbour has destroyed the second, which has also started to burn. After we have moved forward another 500 metres, about forty enemy tanks appear on the horizon. They advance past the blazing wrecks of the first two we have destroyed then stop, shoot, then move forward again, this time firing quickly on the move, one shell after another. Again, rounds splatter against our frontal plate and hull but do us no harm. The tank battle has now begun in earnest.

CREIGHTON 97

A *Tiger* of *Das Reich* moves through a landscape of the type described in the correspondent's account of 8 July.

… the two sides face each other on slopes 1,000 metes apart. Both sides want to sway fate with their shooting. All of the Tigers are firing now. The fighting rises to a climax, but the men who drive and operate these machines must remain clam. They aim quickly, load quickly and respond to orders quickly. We move forward a few metres: move to the right and then to the left, manoeuvring ourselves outside of the enemy's sights. He however comes into the cross-hairs of our own and we fire. We count the burning torches of enemy tanks that will never again fire on German soldiers. After one hour, twelve T-34s are ablaze. The remaining thirty are circling back and forth firing off their shells in a bid to hit us. They shoot well but our armoured plate is strong. We no longer wince when a shell hits our tank. We wipe off the chipped slivers of paint from our faces, reload, aim and shoot. The battle lasted for four hours.

Over the course of the next few days the Germans continued to advance, albeit far more slowly than pre-battle planning had allowed for; Soviet resistance was ferocious. On the fourth day of the offensive, a virtuoso performance by the crew of a damaged Tiger under the command of Unterscharführer Staudeggar manoeuvred the Tiger to maximise the advantage of his optics and long-range gun to target the oncoming T-34s, destroying 17 before the Soviets withdrew out of range to regroup. Another charge saw the Tiger defeat a further five T-34s. The loss of 22 tanks in short order was enough to convince the

Tiger IE of the 1st SS pz.Gren.Div Leibstandarte Adolf Hitler. During Operation Zitadelle the armour of this regiment carried a white bar rather than conventional insignia. They employed an atypical turrent numeral system. The '13' denotes the company number, the '08' the Tiger number.

CREIGHTON 97

Soviets to turn tail, and gain Staudeggar the Knight's Cross.

On 12 July, the SS Panzer Corps was involved in what has since become the almost mythic tank battle at Prokhorovka. During the course of the day the three divisions, albeit with the heaviest Russian assaults being targeted at the LAH in the centre of the line, found themselves having to weather waves of T-34s.

Although this tank battle has come to be linked inextricably with the Tiger with the Russians claiming that they fought against very large numbers of the machines, German records show that there were actually few serviceable Tigers in action that day. Totenkopf was fielding the largest number with 10, Das Reich either one or none and the Leibstandarte – which was facing the heaviest Soviet assaults – just four. The latter were under the temporary command of possibly the most famous tank commander of the Second World War, Michael Wittmann. He and his crew had already accounted for a sizeable number of enemy tanks when he led out the four Tigers of the heavy company at about 0600 hours.

Did you know?

Starting in September 1943 all German tanks, including the Tiger I and the Tiger II, had a concrete-like material called *Zimmerit* applied in a rippled fashion to all vertical surfaces; it was designed to prevent magnetic mines adhering to the tank. It is possible to place the production dates for Tiger Is and IIs if they carry this finish.

The Soviet Fifth Guards Tank Army was attacking en masse and at speed in the desperate hope that they could close the range gap on the long-ranging German tank guns before the latter could respond and exploit that advantage to the full. The Tigers were at the head of the German armoured force and halted on a low rise. At least 100 Soviet tanks appeared and the Germans began firing on them from 1,800m. Many Soviet tanks were hit, but the main body of the enemy tanks, the 181st Brigade of the Soviet 18th Tank Corps, continued to come. The Soviets were trying to close the distance as quickly as possible because they knew they had to get within 500 metres to pose a threat to the heavily armoured Tigers. Wittmann's crew, and those of Loetzsch and Hoeflinger, maintained a high rate of fire. By the time the Soviet tanks were within 1,000 metres, every shot was a direct hit.

By day's end the Tiger company and the IInd Abteilung SS Panzerregiment I had defeated 163 Russian tanks. One Tiger was knocked out with a further nine under repair. All ten of Totenkopf's Tigers were unserviceable with battle damage on the 13th. By the time all offensive action was called off in the south of the salient four days later, 4th Panzer Army had suffered just three total Tiger losses – one from each of the three SS divisions.

ARMY DETACHMENT KEMPF

Although s.Pz.Abt 503 was at full strength with 45 Tigers at the start of Citadel, albeit with 42 actually serviceable on opening day of the offensive, the battalion had been split up with one each of its companies being allocated to the three armoured divisions of the 3rd Panzer Corps. Even though the battalion commander had been vociferous in his opposition to this order he had been overruled.

On the first day of the offensive, the second company ran into an unmarked minefield. From 7 July, all three companies were involved in heavy fighting supporting their respective panzer divisions, and it was not until the 10th that the battalion was back together with 22 operational Tigers. Three Tigers became total losses between the 10th and the 14th. In the

▼ Tiger IE of the 1st company s.Pz.Abt 503 during Operation Zitadelle, July 1943.

Tank combat on the Eastern Front, summer 1944. Three Tigers advance having just destroyed a T-34 and other Red Army vehicles.

In mid-May 1944, SS Pz.Abt 101 was photographed as it undertook a field exercise near Amiens in Northern France. This Tiger IE is from the 3rd Company.

period between the start of Citadel and 21 September, this battalion claimed 501 Soviet tanks destroyed for the loss of 18 Tigers, while the repair crews returned 240 Tigers to service.

On 6 June 1944, the s.SS Pz.Abt 101 was stationed many miles away from Normandy when the Allied landings took place. It started to move in the very early hours of the following day but the first and second companies did not finally arrive until 12 June because of the effect of Allied air attacks.

In consequence of many breakdowns and air attacks en route, just 14 Tigers were on hand. The first company was deployed some miles to the north of the second, which was under the command of *Obersturmfuehrer* Michael Wittman. Having been forced to move several times because of heavy naval gun fire, Wittman's Tigers were now in position near Reference Point 213 just to the north-east of the small town Villers-Bocage. He had been charged with defending the western flank of the 1st SS Panzer Corps although there was no immediate expectation of an Allied attack. So there was much consternation when

◀ *Mid-production Tiger I33, of SS s.Pz.Abt 101 under the command of Oberscharführer Zahner moves through a French village of Morgny on 7 June, en route to the invasion front.*

at 0800 hours Wittmann, who was in his command post, was told that a large enemy column was approaching along the Caen-Villers-Bocage road. He was later to observe that he 'saw tanks rolling past about 150 to 200 metres distant. They were English and American types. At the same time I saw that the tanks were accompanied by armoured personnel carriers.' What Wittman was witnessing was the advance of the 22nd Armoured Brigade – a battlegroup of the British 7th Armoured Division – the famous 'Desert Rats'.

What transpired has since become the stuff of legend and certainly ranks as one of the most remarkable actions ever

➤ *Late model Tiger IE from the SS s.Pz.Abt 101 as used in Normandy from 12 June 1944.*

undertaken by a single tank. As the rest of the company was not yet ready for action, Wittmann jumped into Tiger '222' and moved forward alone. Opening fire, he first knocked out a Cromwell and Sherman Firefly. He then ordered the driver to run the Tiger parallel to the road and alongside the Allied column which he then proceeded to demolish. As his Tiger advanced, Wittmann destroyed a succession of vehicles. Firing his 88mm gun at point blank range he knocked out 13 M3 halftracks, 3 Stuart light tanks, 2 Sherman observation tanks, the M3 of the medical officer and more than a dozen Bren and Lloyd carriers towing 6-pounder anti-tank guns. By this time '222' had reached the edge of Villers-Bocage where it proceeded to knock out a further three of four Cromwell tanks. Turning into the town, the Tiger then drove forward but was stopped by fire from a number of British tanks. Turning around, Wittmann then drove his charge back the way he had come, in the process despatching the fourth Cromwell that had been attempting to stalk the Tiger. Ample testimony to the effect of the Tiger's armour can be gauged by the fact that the Cromwell had managed to get off two AP shots at just 50m but they failed to penetrate the panzer's thick hide!

Shortly thereafter, '222' broke down and Wittmann and his crew abandoned it in the main street. They escaped across country and made contact with elements of the Panzer Lehr Division and Wittman returned with them and a detachment of 15 Panzer IVs. In the meantime, the other Tigers joined the fray and were later joined by those of the 1st Company. Fierce fighting

➤ Early production Tiger IIs of s.Pz.Abt 503 with the P-turm at the Mailly le Camp training ground in France prior to being despatched to the Normandy Front, August 1944.

➤➤ One of the very few Tiger Is to escape destruction in Normandy. Tiger '213' – a late model – made it as far as the river Seine only to find all the bridges had been bombed. Shortly after this picture was taken, the crew destroyed the tank.

continued in and around Villers-Bocage for much of the rest of the day. At its end, the Tiger companies had lost three machines and the British 26 Shermans and Cromwells. Some 230 British prisoners were taken from the column shot up by Wittmann.

➤ This Henschel Turm Tiger II of s.Pz.Abt 503 is seen in Budapest in October 1944. It is covered in Zimmerit so was produced after September 1943. The sheer bulk of the machine is apparent.

Tiger II '105' is representative of the heavy panzers employed in the Ardennes offensive of December/January 1945. Better known in the US as the Battle of the Bulge.

Tiger I E of the Panzer Division Muncheberg, Berlin, May 1945.

For this action, he was awarded the Knight's Cross with Swords. Wittmann died on 8 August when his Tiger was destroyed by Canadian Shermans. By that date the fate of the Germans in Normandy was a foregone conclusion. Very few escaped the destruction of the army of the west.

ENDGAME

Tiger Is, IIs and Jagdtigers were encountered by the Allies and the Red Army, albeit in declining numbers, through to the end of the war. A number were employed in the Ardennes offensive in December and during the invasion of Germany itself. One of the very last Tiger Is was destroyed by the Red Army near the site of the Reichstag in Berlin in May 1945, just days before the German surrender. The guns of the few survivors are now silent but the legend of the Tiger tanks lives on!

Tiger II of SS s.Pz.Abt 502, Germany 1945.

Unit	Number of tanks destroyed
s.Pz.Abt. 501	150(Africa), + 300 (E.Fr)
s.Pz.Abt. 502	+1400 tanks +2000 guns (E.Fr)
s.Pz.Abt. 503 *	+1700 tanks +2000 guns(W/E.Fr)
s.Pz.Abt. 504	150 (Afr) & +100 (It)
s.Pz.Abt. 505	+900 tanks & +1000 guns (E.Fr)
s.Pz.Abt. 506	+400 tanks (W/E.Fr)
s.Pz.Abt. 507	+ 600 tanks (E.Fr)
s.Pz.Abt. 508	+100 Tanks (It)
s.Pz.Abt. 509	+ 500 tanks (E.Fr)
s.Pz.Abt. 510	+ 200 tanks (E.Fr)
Gross Deutschland – Kp and Abt	+ 600 tanks
SS s.Pz.Kp 1	+ 400 tanks (E.Fr)
SS s.Pz.Kp 2	+ 250 tanks (E.Fr)
SS s.Pz.Kp 3	+ 500 tanks (E.Fr)
SS s.Pz.Abt. 101 (501)	+ 500 tanks (W/E.Fr)
SS s.Pz.Abt. 102 (502)	+ 600 tanks (W/E.Fr)
SS s.Pz.Abt. 103 (503)	+ 500 tanks (E.Fr)

* s.Pz.Abt 503 was the most successful of all Tiger battalions. It also had highest scoring tiger commander – Kurt Knispel with 168 kills.

Name	Kills	Unit
Kurt Knispel	168	s.Pz.Abt. 503
Martin Schroif	161	s.SS-Pz.Abt. 1-2
Otto Carius	150+	s.Pz.Abt. 502
Johannes (Hans) Bolter	139+	s.Pz.Abt. 502
Michael Wittmann	138	s.SS-Pz.Abt. 101
Paul Egger	113	s.SS-Pz.Abt. 102
Arno Giesen	111	8./s.SS-Pz.Rgt. 2
Heinz Rondorf	106	s.Pz.Abt. 503
Heinz Gartner	103	s.Pz.Abt. 503
Wilhelm Knauth	101+	s.Pz.Abt. 505
Albert Kerscher	100+	s.Pz.Abt. 502
Balthazar (Bobby) Woll	100+	s.SS-Pz.Abt. 101
Karl Mobius	100+	s.SS-Pz.Abt. 101
Helmut Wendorff	95	s.SS-Pz.Abt. 101
Will Fey	80+	s.SS-Pz.Abt. 102

Eric Litztke	76	s.Pz.Abt. 509
Emil Seibold	69	s.SS-Pz.Abt. 502
Karl Brommann	66	s.SS-Pz.Abt. 503
Alfred Rubbel	60+	s.Pz.Abt. 503
Konrad Weinert	59	s.Pz.Abt. 503
Walter Junge	57+	s.Pz.Abt. 503
Bobby Warmbrunn	57	s.SS-Pz.Abt. 101
Jurgen Brant	57	s.SS-Pz.Abt. 101
Heinz Kling	51+	s.SS-Pz.Abt. 101
Heinz Kramer	50+	s.Pz.Abt. 502
Alfredo Carpaneto	50+	s.Pz.Abt. 502
Heinz Mausberg	50+	s.Pz.Abt. 505
Oskar Geiner	50+	s.SS-Pz.Abt. 103
Johann Muller	50+	s.Pz.Abt. 502
Joachim Scholl	42	s.SS-Pz.Abt. 102
Franz Staudegger	35+	s.SS-Pz.Abt. 101

KURT KNISPEL

MARTIN SCHROIF

OTTO CARIUS   **JOHANNES (HANS) BOLTER**

MICHAEL WITTMANN

PAUL EGGER

HEINZ RONDORF

HEINZ GARTNER

WILHELM KNAUTH

ALBERT KERSCHER

BALTHAZAR (BOBBY) WOLL

KARL MOBIUS

HELMUT WENDORFF

WILL FEY

ERIC LITZTKE

EMIL SEIBOLD

KARL BROMMANN

ALFRED RUBBEL

KONRAD WEINERT

WALTER JUNGE

HEINZ KLING **HEINZ KRAMER**

ALFREDO CARPANETO

HEINZ MAUSBERG

FRANZ STAUDEGGER

TIGER I

Weight:	56.9 tonnes
Length:	6.29m (20ft 8 in)
	8.45m (27ft 9 in) (gun forward)
Width:	3.55m (11ft 8 in)
Height:	3.0m (9ft 10 in)
Crew:	5
Armour:	25–120mm (0.98–4.7in)
Primary armament:	1× 88mm KwK 36 L/56
	(92 rounds)
	(106 and 120 rounds for some modifications)
Secondary armament:	2×7.92mm Maschinengewehr 34
	(4,800 rounds)
Engine:	Maybach HL230 P45
	700 PS (690.4hp, 514.8 kW)
Suspension:	torsion bar
Operational range:	110–195km (68–120 miles)
Speed:	38kmph (24mph)

➤ *A Tiger IE of the GD, Autumn 1943.*

121

TIGER II

Weight:	68.5 tonnes (early turret)
	69.8 tonnes (production turret)
Length:	6.4m (21ft 0in)
	10.286m (33ft 9in) (gun forward)
Width:	3.755m (12ft 4in)
Height:	3.09m (10ft 2in)
Crew:	5
Armour:	25–180mm (1–7in)
Primary armament:	1×88mm KwK 43 L/71
	(Porsche turret: 80 rounds)
	(Production turret: 86 rounds)
Secondary armament:	2× 7.92mm Maschinengewehr 34
	(5,850 rounds)
Engine:	V-12 Maybach HL230 P30
	700 PS (690hp, 515 kW)
Suspension:	torsion bar
Operational range:	120–170km (75–110 miles)
Speed:	41.5 kmph (25.8mph)

1942

22 September: Four Tiger Is of s.Pz.Abt 502 are committed to action in support of German 170th Infantry Division operating to the south of Leningrad.

1942

10 November: Hitler orders s.Pz.Abt 501 to be shipped to North Africa.

1942

23 November: Tigers of 1/501 arrive at Bizerte.

1942

29 November: Tigers of 1/501 fight at the Battle of Tebourba against US armoured units.

1943

January: Hitler insists that the armour on the Tiger II be strengthened. That on the hull front is raised to 150mm and the sides to 80mm.

February: All new AFVs to receive a base coat of sand yellow paint, with dark green and red brown over-sprayed to suit local conditions.

March: Start of production of the Tiger variant the *Befehlswagen* (PzBefWg Tiger).

16 March: The new heavy tank Tiger H3 becomes known as the Tiger II.

10 May: Pz.Abt 503 is deployed to Kharkov.

July: Decision taken to raise a SS heavy tank battalion for service with the newly created 1 SS Panzer Corps. It is to have 45 Tiger Is.

July: Operation Citadel (*Unternehmen Zitadelle*) is launched. The largest number of Tiger Is ever assembled for one operation is employed at the Battle of Kursk in Russia.

20 October: a wooden mockup of the *Jagdtiger* is constructed on a Tiger II chassis and presented to Hitler.

December: Tiger II goes into production at the Kassel works alongside the Tiger I.

January: First Tiger II leaves the Kassel works.

February: Two prototypes of the *Jagdtiger* produced.

1944

May: Tiger II sees action for the first time near Minsk.

1944

June 13: At the battle of Villers-Bocage, a week after D-Day, SS-Hauptsturmführer Michael Wittman (with minimal help from his platoon) catches the leading elements of the British 22nd Armoured Brigade unawares and his Tiger I destroys more than two dozen Allied vehicles, including several tanks, in 15 minutes.

1944

July: First *Jagdtiger* leaves the factory.

1944

August: The last Tiger I leaves the production line.

1944

September: Start of bombing raids on the Kassel works, which continue until early October and destroy most of the plant.

7 September: Tiger I gets its final official denomination of Panzerkampwagen Tiger Ausf.E (Sd.kfz.181. Sd.Kfz stands for *Sonderkraftfahrzeug* (special purpose vehicle, or special ordnance vehicle).

March: The last Tiger II leaves the production line.

May: The last *Jagdtiger* leaves the production line.

May 10: A Tiger II of *schwere Panzer Abteilung* 503 is the last German tank destroyed in the war, blown up by its crew in Austria.

BIBLIOGRAPHY

Healy, Mark, *Zitadelle: the German Offensive against the Kursk Salient 4–17th July, 1943*, The History Press

Jentz, Thomas and Hilary Doyle, *Germany's Tiger Tanks Volume 1: D.W. to Tiger 1: Design, Production and Modifications* ,Schiffer Books

Jentz, Thomas and Hilary Doyle, *Germany's Tiger Tanks Volume 2: VK 45.02 to Tiger II: Design , Production and Modifications* ,Schiffer Books

Jentz, Thomas and Hilary Doyle, *Germany's Tiger Tanks Volume 3: Tiger I and II: Combat Tactics* ,Schiffer Books

Jentz, Thomas, Hilary Doyle and Peter Sarson, Tiger I Heavy Tank 1942–45, Osprey Publishing

Jentz, Thomas, Hilary Doyle, Peter Sarson and Lee Johnson, Kingtiger Heavy Tank 1942–45, Osprey Publishing

Kleine, Egon and Volkmar Kuhn, Tiger: *The History of a Legendary Weapon 1942–45*, J.J. Fedorowicz Publishing

Schneider, Wolfgang, *Tigers in Combat Volume 1 – The Army Battalions*, Stackpole Books

Schneider, Wolfgang, *Tigers in Combat Volume 2 – GD and the Waffen SS Tiger Units*, Stackpole Books

Stansell, Patrick, *The Modeler's Guide to the Tiger Tank*, Ampersand Publishing